FLORA OF TROPICAL EAST AFRICA

HAMAMELIDACEAE

B. Verdcourt

Evergreen or deciduous trees or shrubs, often with stellate indumentum. Leaves alternate or less often opposite, simple, but often pinnatifid or palmatilobed, entire or serrate (the teeth sometimes glandular), usually stipulate; stipules mostly paired, persistent or more usually deciduous. Inflorescences terminal or axillary, racemose, often spicate or capitate, sometimes so densely packed that the flowers are almost joined; bracts and bracteoles often present, the former sometimes forming a coloured involucre. Flowers small to large, regular or rarely irregular, hermaphrodite or unisexual, hypogynous to perigynous, rarely without a perianth. Calyx-tube variously shaped, usually ± adnate to the ovary; lobes (3–)5(–7), imbricate or valvate. Petals 4–5 or more, sometimes absent in ♀ flowers, free, imbricate or valvate, rarely circinate, mostly linear-spathulate or obovate. Stamens 4–5(–25), rarely fewer, as many as and alternating with the petals, arranged in one series; staminodes sometimes present; filaments free, sometimes thickened and shorter than the anthers; anthers 2-thecous, opening lengthwise or by valves, the connective sometimes projecting to form a beak. Disc usually absent, when present annular or composed of separate glands. Ovary inferior, half-inferior or rarely superior, (1–)2(–3)-locular, the carpels sometimes free at the apex; ovules 1 or more in each cell, apical or anatropous, pendulous; placentation axile; styles 2, subulate, usually free, often recurved, sometimes persistent, the stigmas terminal or lateral. Fruit a woody loculicidal or septicidal capsule, with leathery exocarp and bony endocarp, often appearing apically 4-valved. Seeds 1–many, sometimes winged, with thin endosperm; embryo straight.

A fairly small family of 19–22 genera and about 50–100 species according to delimitation, mostly in subtropical regions and mostly in the N. hemisphere; one genus only in continental tropical Africa.

TRICHOCLADUS

Pers., Syn. 2: 597 (1807); Hutch. in K.B. 1933: 427–430 (1933);
Breitenbach, Indig. Trees S. Afr.: 240 (1965)

Trees or shrubs, mostly with conspicuous stellate indumentum. Leaves alternate or opposite, simple, usually entire but ± toothed in one species; stipules small. Flowers in dense terminal or axillary round heads, ♂ or sometimes only ♀. Calyx-tube shallow or campanulate, sometimes adnate to the ovary; lobes (3–)4–5, broadly triangular, valvate. Petals 4–5 or often absent in ♀ flowers, linear-spathulate, much longer than the calyx, often free at base, valvate or slightly imbricate, at first sometimes rolled up or folded in the bud, later sometimes with revolute margins. Stamens as many as the petals; filaments usually shorter than the anthers, turbinate or obpyramidal; anthers opening by valves, mostly beaked; staminodes absent. Ovary inferior to almost superior, 2-locular; ovules pendulous, solitary in each cell. Capsule subglobose, (1–)2-locular, 2-valved, but appearing 4-valved at apex after dehiscence. Seeds ellipsoid or ovoid-oblong, the hilum divided into 2 areas extending from proximal end of the seed on each side; embryo large, the cotyledons well developed.

A small genus of 5 species confined to Africa and extending from South Africa up through East Africa to Ethiopia; three species in the Flora area, two of which are still very poorly known.

Considerably more field work needs to be done to elucidate variation in the sex of the flowers. Various works state that the flowers can be monoecious or dioecious and that the petal-size varies in the differently sexed flowers. I have found anthers and ovules in all the long-petaled and short-petaled flowers I have examined and suspect that the rapid lengthening of the petals during anthesis has been misunderstood so far as the East African species are concerned. I have found no buds which do not have petals.

Leaves alternate, persistently covered with fine greyish to
      coppery stellate tomentum beneath, never toothed nor
      minutely peltate at the base   .   .   .   .   1. *T. ellipticus*
Leaves opposite, at length glabrous but when young covered
      with coarser reddish-brown stellate hairs:
      Leaves entire, not in any way peltate or with but faintest
          tendency to be so   .   .   .   .   .   2. *T. goetzei*
      Leaves slightly to distinctly dentate, mostly very slightly
          peltate at the base   .   .   .   .   .   3. *T. dentatus*

1. **T. ellipticus** *Eckl. & Zeyh.*, Enum. Pl. Afr. Austr.: 356 (1837); Sond. in Fl. Cap. 2: 325 (1862); V.E. 3(1): 290 (1915); Harms in E. & P. Pf., ed. 2, 18a: 323 (1930); Hutch. in K.B. 1933: 428 (1933). Type: South Africa, Cape Province, Albany, Bosjesmansrivier near Philipstown and Balfour, *Ecklon & Zeyher* 2270 (B, holo. †)

Evergreen shrub or tree, sometimes somewhat scrambling, (2–)5–10(–18) m. tall; bark greyish-white or creamy-brown, smooth or somewhat rough; young branchlets tomentose with fine ochraceous or ferruginous stellate hairs, older ones glabrescent. Leaves alternate; blades elliptic, obovate-elliptic to oblanceolate, (1·2–)5–28 cm. long, 0·8–12 cm. wide, acute to acuminate, cuneate to rounded or rarely subcordate* at the base, entire, mostly rather coriaceous, glabrous above, densely persistently stellate tomentose beneath with fine greyish, silvery, yellowish-brown or pale coppery hairs, frequently with scattered darker rusty spots; petiole 0·5–1·8 cm. long; stipules linear, soon falling. Inflorescences capitate, axillary or appearing terminal, 0·7–2 cm. in diameter; peduncle 0·5–1·5 cm. long; bracteoles filiform, 2 mm. long. Flowers sessile, tomentose, sweetly scented (or unpleasantly *fide Greenway* 8544). Hermaphrodite flowers: true calyx-tube 1 mm. long, tomentose; limb 2 mm. long, its rounded lobes 1 mm. long, recurved; petals white, greenish-yellow or yellow, 0·9–1·7 cm. long, 1·7–2 mm. wide, revolute, tomentose on outer surface; stamens 5, the filaments obpyramidal, 0·8–1·2 mm. long, 0·6 mm. wide; anthers 1–1·2 mm. long; connective prolonged into an erect beak; styles 1·3 mm. long; ♀ flowers stated to be similar but with petals reduced, 4 mm. long, or even lacking (see introductory note after generic description). Capsule subglobose, 6–7(–8) mm. long, 5 mm. across, the apical valves conspicuous, pubescent, 1–2-seeded. Seeds yellowish or greyish, mottled black, smooth, ovoid or ellipsoid, flattened, 5 mm. long, 3–4 mm. wide; hilar area brownish, 3–4 mm. long situated across the end of the seed; on one side, within this area, there is also a small elliptic patch, the true hilum, 1·2 mm. long.

subsp. **malosanus** (*Bak.*) *Verdc.* in K.B. 24: 345 (1970). Type: Malawi, Mt. Malos a *Whyte* (K, holo. !)

* Sometimes marked in juvenile foliage of Ethiopian variants.

FIG. 1. *TRICHOCLADUS ELLIPTICUS* subsp. *MALOSANUS*—1, fertile branch, × 1; 2, stellate hair, × 12; 3, flower, × 4; 4, calyx and corolla opened out to show stamens, × 4; 5, petal, × 4; 6, stamen, front view, × 6; 7, same, side view, × 6; 8, longitudinal section of gynoecium, × 9; 9, styles × 6; 10, infructescence, × 1; 11, dehisced capsule, × 2; 12, seed, × 2. 1, 3–9, from *C. G. Rogers* 5; 2, 10–12, from *A. S. Thomas* 3617.

Leaf-blades more often broadly elliptic, mostly larger, at least 7–10 cm. long, 4–6 cm. wide. Inflorescences usually larger than in the typical subspecies. Fig. 1.

UGANDA. Acholi District: SE. Imatong Mts., Aringa R. head-waters, 6 Apr. 1945, *Greenway & Hummel* 7297!; Mbale District: Suam R., Oct. 1930, *Brasnett* 36!; Mengo District: Busiro-Kyadondo boundary, Kajansi Forest, 26 Aug. 1949, *Dawkins* 345!

KENYA. Ravine District: Timboroa to Mau Summit, 15 Sept. 1958, *Napper* 829!; Machakos District: R. Athi at Fourteen Falls, 26 May 1963, *Verdcourt* 3632!; Masai District: Mara Bridge, 15 Apr. 1961, *Glover et al.* 543!

TANGANYIKA. Moshi District: Rongai, Jan. 1956, *Carmichael* 544!; Lushoto District: Shume, 7 Jan. 1947, *Hughes* 15!; Kilosa District: summit of Mt. Luemba, 15 Feb. 1933, *B. D. Burtt* 4554!

DISTR. U1–4; K1–6; T1–3, 6, 7; Congo, Ethiopia, Malawi, Mozambique, Zambia and Angola

HAB. Upland rain-forest, often dominant by streams and in swampy places, also extending into riverine forest, *Acacia lahai* woodland and swamp forest;1140–2700m.*

SYN. *T. malosanus* Bak. in K.B. 1897: 266 (1897); Hutch. in K.B. 1933: 430 (1933); T.T.C.L.: 244 (1949); I.T.U., ed. 2: 155, t. 6 (1952); F.F.N.R.: 67 (1962)
   T. ellipticus Eckl. & Zeyh. var. *latifolius* Engl. in E.J. 49: 456, fig. 1/H–N (1913),** *nom. non rite publ.*
   [*T. ellipticus* sensu V.E. 3 (1), t. 188/H–N (1915); Harms in E. & P. Pf., ed. 2 18a, fig. 169/H–N (1930); Germain in F.C.B. 2: 582 (1951); Jex-Blake, Gard. in E. Afr., ed. 4, t. 15 (1957); K.T.S.: 233, t. 13 (1961); Burger, Fam. Fl. Pl. Eth.: 85, fig. 15/4 (1967); Teixeira in Bol. Soc. Brot., sér. 2, 43: 162, t. 1–4 (1969), *non* Eckl. & Zeyh. sensu stricto]

DISTR. (of species as a whole). Typical *T. ellipticus* with usually more lanceolate, smaller, often more acuminate leaves occurs in South Africa

NOTE (on species as a whole). As Germain mentions, the form and dimensions of the leaves in this species are exceedingly variable even on one branch, but comparing the whole South African material with all the tropical material available there is a distinct difference in average size and facies. Occasional East African specimens do match South African material fairly closely, e.g. *Semsei* 2471! (Tanganyika, Njombe District, Elton Plateau, R. Lumakarya, 7 Sept. 1956) has small leaves and is indistinguishable from certain Natal specimens which are however not typical of subsp. *ellipticus*. *Greenway* 6529! (Pare District, Mtonto, 4 July 1942) is an evergreen tree to 27 m. and has more narrowly lanceolate leaves but does not seem to be specifically distinct; unfortunately the material is sterile.

2. **T. goetzei** *Engl.* in E.J. 49: 455, fig. 1/A–G (1913) & V.E. 3(1), t. 188/A–G (1915); Harms in E. & P. Pf., ed. 2, 18a: 323, fig. 169/A–G (1930); T.T.C.L.: 244 (1949). Type: Tanganyika, Iringa District, Uhehe, Uzungwa [Utschungwe] Mts., ? near Kissinga, *Goetze* 574 (B, holo. †)

Small evergreen tree 6–10·5(–24, *fide Ede* 2) m. tall; bark purple-brown, rather rough, flaking in oblong patches; young branchlets very densely hairy with rather coarse reddish-brown stellate hairs, older ones glabrescent. Leaves opposite; blades elliptic, elliptic-lanceolate or oblong, 3–11 cm. long, 0·8–4·2 cm. wide, rather rounded to acute or obtusely acuminate at the apex, truncate to cuneate at the base, not or only minutely peltate (one such leaf observed), entire, the margins sometimes revolute, rather coriaceous, at first thickly covered with rather coarse reddish-brown stellate hairs on both surfaces, soon glabrous above and at length glabrous beneath or retaining only a very few scattered hairs on the midrib and nerves; venation rather prominently reticulate on both surfaces; petiole 0·4–2 cm. long; stipules linear, 5 mm. long, soon falling. Inflorescences capitate, axillary and terminal, 2·5–3·5 cm. in diameter; peduncle 0·1–2 cm. long; bracteoles 2·5–3 mm. long, occasionally adnate to the calyx. Flowers sessile, somewhat

* *Fyffe* 15 (Uganda, Bwamba, Oct. 1925) is labelled 2300', but this seems very low and does not agree with the present position of Bwamba.
** The figure given by Engler as representing var. *latifolius* is reproduced in V.E. and E. & P. Pf. as *ellipticus* without any varietal designation.

sunk into the rhachis, tomentose. Hermaphrodite flowers : calyx-tube 1 mm. long; limb 5 mm. long, divided into 3 lobes (or sometimes 4 if an additional small one included), erect, 2·5–3 mm. long and wide, stellate tomentose outside, tipped with a hairy appendage; petals 5–6, white, linear, (1·4–)1·8–1·9 cm. long, (1–)2–3·5 mm. wide, narrowed at the base, not revolute, glabrous; stamens 5–6; filaments oblong, flattened, 1 mm. long; anthers 1·3–1·6 mm. long; connective prolonged into an erect beak; styles 1·3 mm. long; ♀ flowers not seen (see note, p. 2). Capsule ± 1–1·1 cm. long including the 4 mm. long apical valves, 0·9–1 cm. across. Seeds creamy to dark brown, ellipsoid or oblong, shiny, 5–6 mm. long, 2–3·5 mm. broad.

TANGANYIKA. Uluguru Mts., S. Uluguru Forest Reserve, edge of Lukwangule Plateau, 17 Mar. 1953, *Drummond & Hemsley* 1655 !; Iringa District : Mufindi, R. Luisenga, 16 Mar. 1962, *Polhill & Paulo* 1779B ! & Mufindi, Kigogo, May 1953, *Eggeling* 6625 !; Rungwe District : Rungwe Crater, 3 Jan. 1912, *Stolz* 1067 !
DISTR. **T**6, 7; not known elsewhere
HAB. Degraded upland rain-forest of *Syzygium*, *Rapanea*, *Ocotea* and *Ternstroemia*, and on ridges with *Syzygium* and *Faurea*, also riverine forest; 1680–2300 m.

NOTE. So little material is available of this species that a number of problems remain. The type has been destroyed and Engler's description is not especially detailed; he gives the petal-width as 1 mm., but in *Drummond & Hemsley* 1655 & *Stolz* 1067, which I have associated with this species, the petals are much wider. I presume Engler's dimensions refer to immature petals.

3. **T. dentatus** *Hutch.* in K.B. 1933 : 430, fig. (1933); T.T.C.L. : 244 (1949). Type : Tanganyika, Ukinga, Mwakaleli [Kwakalila], *Stolz* 2297 (K, holo. !, BM, EA, iso. !)

Small shrubby tree about 4 m. tall; young branchlets densely tomentose with coarse ferruginous stellate hairs, compressed at the nodes. Leaves opposite; blades broadly oblong-lanceolate or elliptic, 8–15 cm. long, 3–6 cm. wide, acuminate, rounded to truncate at the base, slightly to distinctly peltate with the petiole inserted up to 7 mm. above the base, coriaceous, obscurely to distinctly repand-dentate, very densely stellate velvety tomentose with ferruginous hairs on both surfaces when young but soon glabrous above and practically glabrous beneath; petiole 1·5–3 cm. long; stipules linear, 6 mm. long, soon falling. Inflorescences capitate, axillary and terminal, 1·5–2·5 cm. in diameter; peduncle 0·5–1·5 cm. long or sometimes appearing longer where the first pair of leaves below the inflorescence have failed to develop; bracteoles up to 7 mm. long. Flowers as in previous species; ♀ flowers not known (see note, p. 2). Capsule not seen.

TANGANYIKA. Rungwe District : Kyimbila, Ukinga, Mwakaleli [Kwakalila], 13 Nov. 1913, *Stolz* 2297 !
DISTR. **T**7 (known only from the type)
HAB. Bamboo forest; about 1500 m.

NOTE. Although the foliage is admittedly somewhat different from that of the few known specimens of *T. goetzei*, the general facies is very similar and I doubt very much if *T. dentatus* is more than a state of that species. The large peltate leaves are strongly suggestive that they might be from a young tree just starting to flower. Even in *T. goetzei* the leaves are sometimes minutely peltate. Equally suspicious is that in *T. crinitus* (Thunb.) Pers. leaves may be peltate or not. Further material is much desired to resolve the problem.

INDEX TO HAMAMELIDACEAE